Handling Data

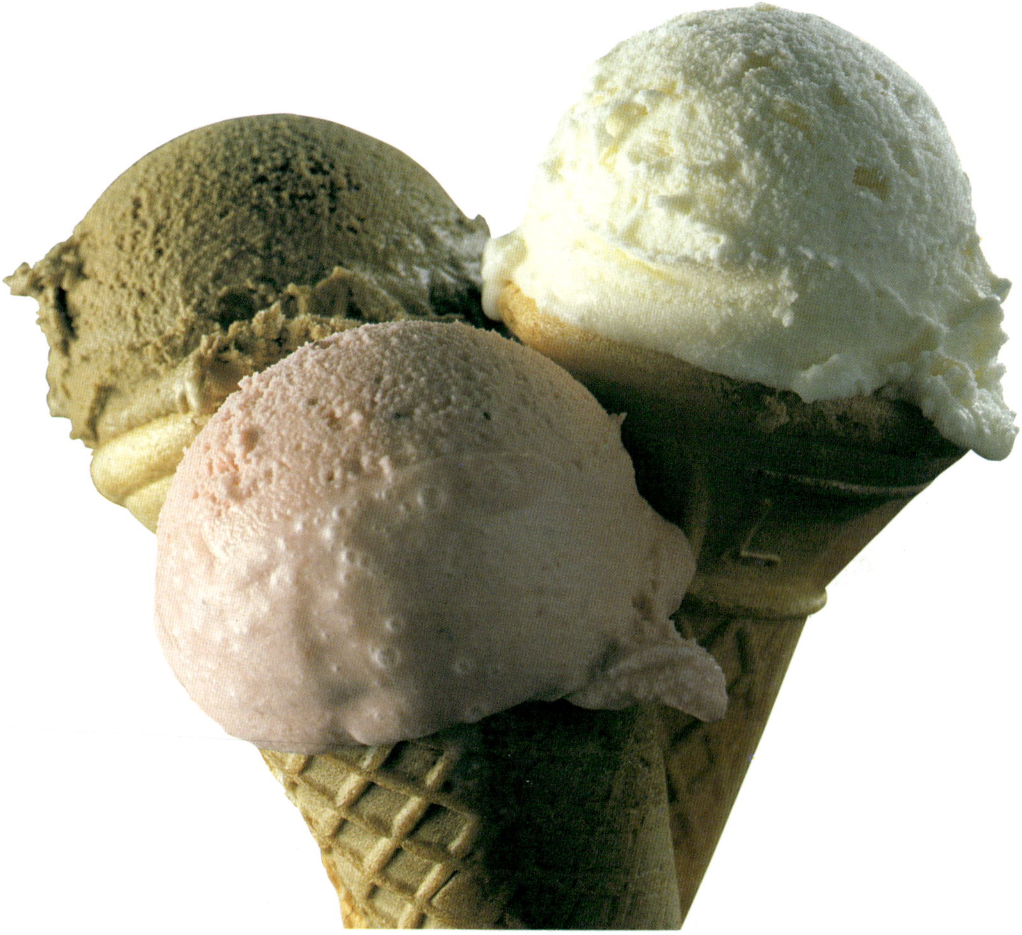

David Kirkby

First published in Great Britain by Heinemann Library
an imprint of Heinemann Publishers (Oxford) Ltd
Halley Court, Jordan Hill, Oxford OX2 8EJ

MADRID ATHENS PARIS
FLORENCE PRAGUE WARSAW
PORTSMOUTH NH CHICAGO SAO PAULO
SINGAPORE TOKYO MELBOURNE AUCKLAND
IBADAN GABORONE JOHANNESBURG

© Heinemann Publishers (Oxford) Ltd

Designed by The Point
Cover design by Pinpoint Design
Printed in China
Produced by Mandarin Offset
99 98 97 96 95
10 9 8 7 6 5 4 3 2 1

ISBN 0431 06894 1

British Library Cataloguing in Publication Data
Kirkby, David
Handling Data. - (Maths Live Series)
I. Title II. Series
510

Acknowledgements
The author and publisher wish to acknowledge, with thanks,
the following photographic sources:
Ace Photo Agency p12; Zefa p20; Anthony Blake p6; Trevor Clifford pp19, 24, 40; Chris Honeywell p38.

The publishers would also like to thank the following for the kind loan of equipment:
NES Arnold Ltd; Polydron International Ltd.

Note to reader: words in **bold** in the text are explained in the glossary on page 44.

CONTENTS

1 DATA

We often have to deal with information and try to sort it out. When we have some facts which give us information about something, we call it **data**.

One way of collecting data is by observation. For example, in a traffic survey, we might watch the traffic and observe the different types of vehicle which are using a road.

Traffic survey

Cars	I I I I
Lorries	I
Bicycles	I
Motor-cycles	
Buses	I I
Others	

As we observe, we will need to record the data, so that it can be remembered and used later. The recording of data is done on a **data collection sheet**.

Another way of collecting data is by measuring. For example, if we are studying trees, we might need to collect data by measuring the lengths of their leaves.

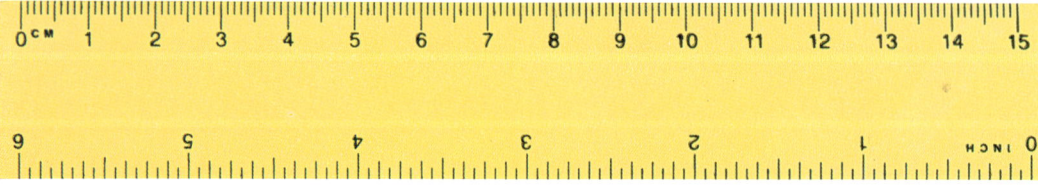

Sometimes data is collected by asking people questions. The questions need to be carefully planned beforehand. The set of questions is called a **questionnaire**.

When the data has been collected and recorded, it is often in a muddle and needs to be tidied up and made simpler. This is called processing the data. Much **data processing** is done by a computer.

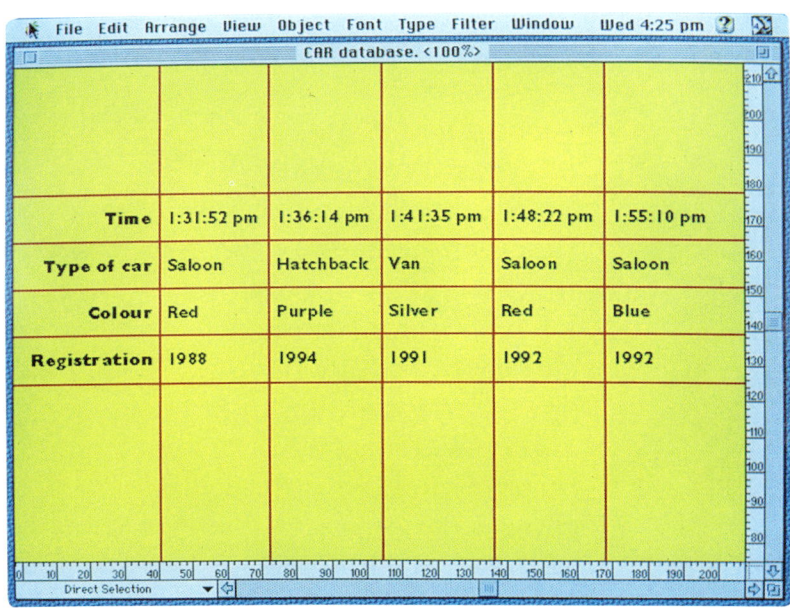

When the data has been processed we look for ways of representing it. **Data representation** means showing it to other people in an easy-to-read way.

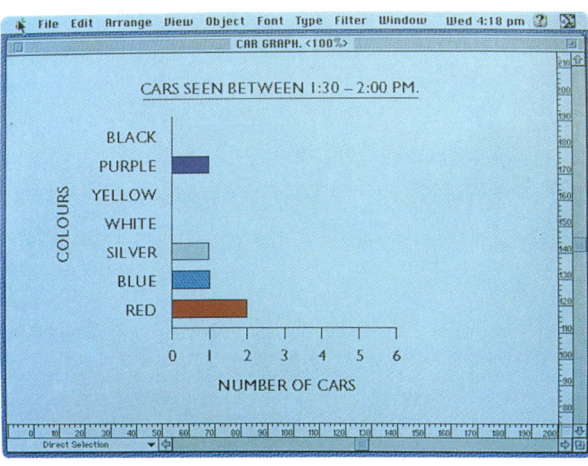

Usually this is by means of graphs, pictures and charts. When we look at the represented data, whether it be our own data or someone else's, we need to be able to 'read' it and understand what it tells us. This is called **interpreting** the data.

In summary there are five stages of handling data. They are collecting data, recording data, processing data, representing data and interpreting data.

Data is all around us – on television, in newspapers, in magazines, in shops – and it is a part of everybody's life both at work and at home. It is therefore important to be skilled in handling it.

2 PICTOGRAPHS

A **pictograph** is a graph which uses pictures to show information.

A shop had a special offer last Wednesday. They offered a free ice-cream to each of the first twenty shoppers. Each shopper could choose from four flavours: strawberry, vanilla, mint and chocolate. The pictograph below shows the chosen flavours.

Choice of flavour of ice-cream

Strawberry	♀ ♀ ♀ ♀
Vanilla	♀ ♀ ♀ ♀ ♀ ♀ ♀ ♀
Mint	♀ ♀ ♀ ♀ ♀
Chocolate	♀ ♀ ♀

Key: ♀ means 1 ice-cream

There are four pictures alongside the strawberry flavour to show that four shoppers chose that flavour.
The **key** shows what each picture means. All pictographs must have a key.

TO DO:

- How many shoppers chose mint flavour?
- Which flavour was most popular?

6

Some children were asked to choose their favourite season.
This pictograph shows the results.

Our favourite season

Spring	👤👤👤👤
Summer	👤👤👤👤👤
Autumn	👤👤👤
Winter	👤👤

Key: 👤 means 2 votes

It is common to draw a pictograph in which a picture
means more than 1. The key explains that one picture
means 2 votes, so that half a picture means 1 vote. The
three and a half pictures alongside Spring show that seven
children voted for this season.

How many voted for each of the other seasons?
How many children voted altogether?

 ## TO DO:

Play the pictograph game

You need a pack of playing cards, shuffled and placed in a
face-down pile.

- Choose two suits each (from
  hearts, clubs, diamonds,
  spades).
- Turn over the top card and
  draw the shape of a card in
  the correct line.
- Continue revealing cards and
  drawing pictures.
- The winner is the person who
  chose the first suit to have five
  pictures.
- Play a variation in which you
  draw half a card each time.

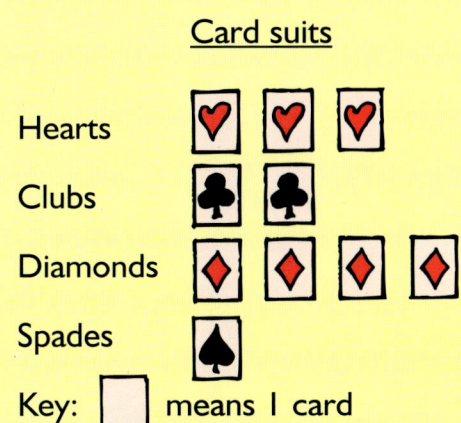

Card suits

Hearts	♥ ♥ ♥
Clubs	♣ ♣
Diamonds	♦ ♦ ♦ ♦
Spades	♠

Key: ☐ means 1 card

3 TALLIES

When some data has been collected, it may need processing. This data shows how many slices of bread were eaten by each of 60 children in one day.

Number of slices of bread eaten yesterday

0	1	3	2	6	4	5	0	4	2
4	2	3	5	4	5	3	4	5	3
5	3	4	5	4	2	6	3	4	4
3	6	5	3	6	0	5	4	6	2
5	4	2	6	4	6	1	4	3	5
6	3	5	0	4	4	5	3	2	4

One way of processing data is to draw a **tally chart**. We make a line for each number and complete a 'gate' to show a group of five lines. These lines are called **tallies**.

Number of slices eaten	Tallies	Total
0	I I I I	4
1	I I	2
2	++++ I I	7
3	++++ ++++ I	11
4	++++ ++++ ++++ I	16
5	++++ ++++ I I	12
6	++++ I I I	8

By counting the tallies, we can easily count the total number of children who ate no slices of bread, one slice of bread, two slices of bread, and so on.

TO DO:

- How many children ate three slices of bread yesterday?
- What was the most common number of slices eaten?
- How many children ate less than three slices?

A group of children found out the ages of all the people in their families. Altogether there were 83 people. Here are their ages.

Ages of people in our families

```
11  30   7  42  28  20  52  45  39  54   8  41  38  25  15
24  32  50   1  59  23  82  36  12  69  21  73   5  49  38
14  42  29  12  30   1  21  75  27  60   4  48  37  53  36
31   6  67  39  21  55  34   2  74  35  58  17  61  23  30
22  53  24  44   3   9  40  26  64  34  65  48   7  31  25
            46  77  20  33  57  13  46  32
```

Because there are so many different ages, it is best to group the data into different age-ranges.

Age-range	Tallies	Frequency
0−9	++++ ++++ I	11
10−19	++++ II	7
20−29	++++ ++++ ++++ I	16
30−39	++++ ++++ ++++ III	18
40−49	++++ ++++ I	11
50−59	++++ IIII	9
60−69	++++ I	6
70−79	IIII	4
80−89	I	1

Notice that each group of ages is the same size – ten years. The word 'total' is sometimes replaced by the word 'frequency'. **Frequency** means 'how many' or 'how often'. When the data is put into groups like this, it is called **grouped data**.

CHALLENGE:

How good are you at reading the grouped data tally chart?
• To which age-range do the most people belong?
• How many age-ranges contain more than ten people?
• How many people are aged 40 or more?
• How many people are aged less than 20?

9

4 BLOCK GRAPHS

A **graph** is a picture which makes data clearer and easier to read and understand. A **block graph** is the simplest type of graph. It is made with a collection of blocks. Each block stands for something.

A block graph can be shown using objects such as cubes. The data below shows the hair colour of a group of children. One cube represents one child with that colour of hair.

Our hair colour

| Black | Brown | Blonde | Red |

A simple way of drawing a block graph is to use gummed squares.

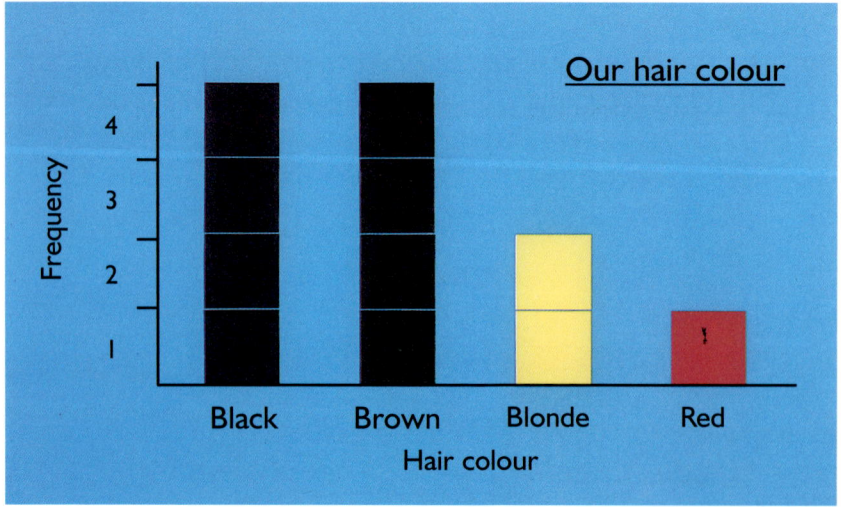

Each square represents one child. Each square is a 'block' on the block graph.

The graph on page 10 has a horizontal and a vertical line which are labelled 'Hair colour' and 'Frequency'. Each line is called an **axis** of the graph. Together they are called **axes**.

The **horizontal axis** shows the hair colour. The **vertical axis** shows the frequency, or the number of children with each hair colour.

In a block graph, the numbering of the frequencies on the vertical axis should be written in the spaces between the blocks.

Notice that the graph has a title. All block graphs should have a title, and both axes should be clearly labelled. There should also be spaces of equal width between each column of blocks.

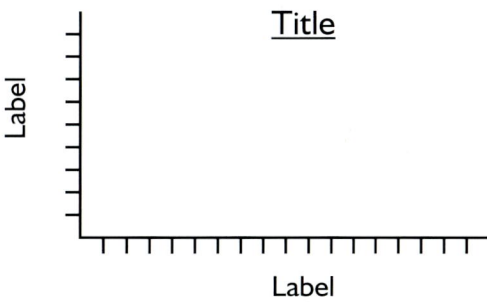

It is important to be able to 'read' or interpret a graph, to understand what the graph shows.

TO DO:

Use the graph on page 10 to answer these questions:
• How many of the children have blonde hair?
• Which hair colour is the most frequent?
• How many more children have black than red hair?
• How many children are in the group altogether?

CHALLENGE:

Look at this block graph.

Can you invent five questions for this graph? Try them out on a friend.

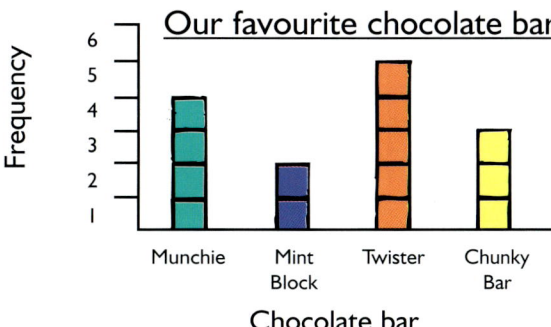

5 BAR GRAPHS

A **bar graph** shows information using bars. It is similar to a block graph, but uses a bar to replace each tower of blocks. Block graphs and bar graphs are often used to show differences between things.

John watched cars travelling past his house and wrote down the colour of each. He watched 40 cars altogether and recorded his data in a tally chart.

Colour of cars passing my house		
Car colour	Tallies	Frequency
Red	⊦⊦⊦ ⊦⊦⊦ ‖‖	13
White	⊦⊦⊦ ‖‖	8
Black	⊦⊦⊦ ⊦⊦⊦	10
Blue	⊦⊦⊦	5
Green	‖‖‖	4

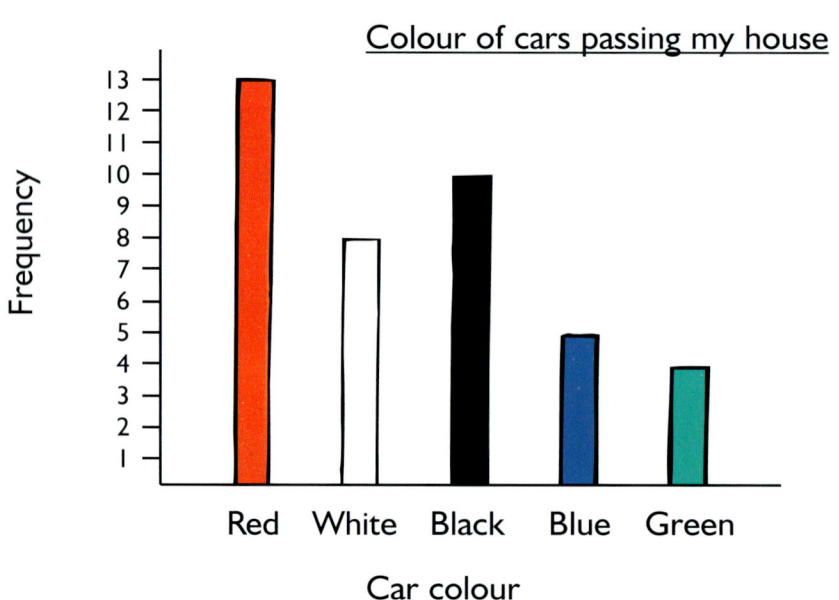

Colour of cars passing my house

As with a block graph, a bar graph must have a title and two labelled axes. The length of the bars shows how many or how much.

The bars should all be the same width, and have gaps of equal width between them.

TO DO:

- Collect your own data on car colour and draw a bar graph to show the results.
- Compare it with the graph on page 12.
- Was red the most popular colour in your survey?

A bar graph is sometimes known as a **bar chart**. For a block graph, the numbering of the frequencies was written between the spaces. For a bar graph, the numbering should be written alongside the axis divisions.

CHALLENGE:

Twenty-three children were asked their favourite kind of biscuit.

- Ginger biscuits were the most popular, with seven votes.
- Digestives were the least popular, with only one vote.
- Twice as many voted for chocolate fingers as voted for shortbread.
- Twice as many voted for wafer biscuits as voted for crackers.

Can you draw the bar graph to show the results?

6 BAR-LINE GRAPHS

A **bar-line graph** is similar to a bar graph except that, instead of drawing a bar to show the frequency, we draw a straight line.

In written texts, some letters are used more often than others. The graph below shows the frequency of each letter used in the passage of writing below left.

Take a closer look at the buttons on your shirt.

Girls' clothes usually button up from the left.

But boys' togs button up from the other side.

Can you guess why?

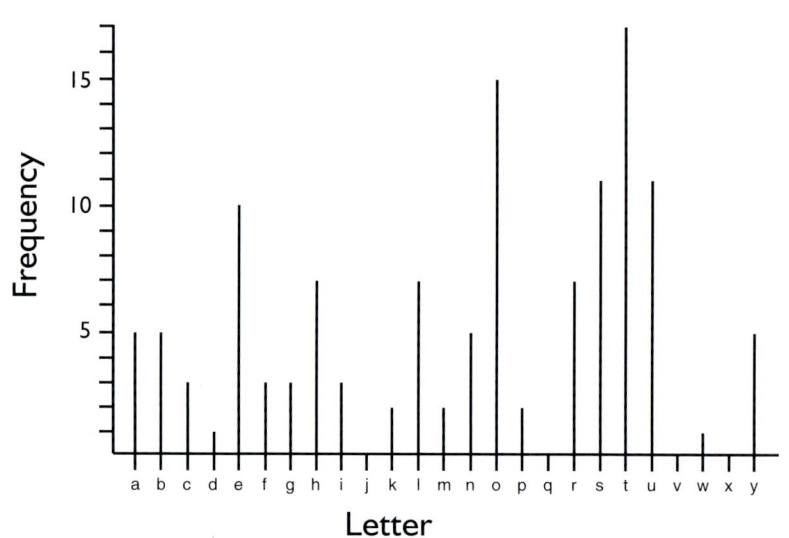

Use of letters in the passage

The lengths of the bars match the frequency with which each letter occurs.

CHALLENGE:

- Choose a book.
- Look at the first 200 words (of two letters or more) in the book.
- Investigate how often each letter of the alphabet is used.
- Look at the scores for each letter in a game of Scrabble.
- What scores are given to the most frequently used letters? Why do you think this is?

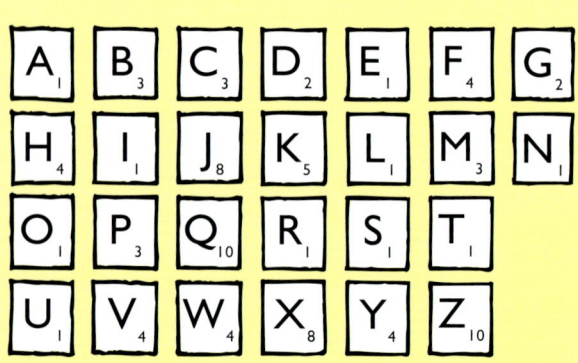

Bar graphs and bar-line graphs are usually drawn so that the bars are **vertical**, but sometimes they are drawn **horizontally**.

Dates of a bag of coins

Coin date / Frequency

Coin date	Frequency
1990	2
1991	4
1992	9
1993	11
1994	7
1995	3

The graph shows that the most common date on the coins is 1993, and the least common is 1990. No coins were found to be more than six years old.

TO DO:

- Collect as many coins as you can and draw a bar-line graph to show the dates of the coins.
- Use squared paper to draw the graph. Start by drawing the axes in the same position as those in the picture above. Sort your coins into piles for each year, count the number in each pile, then draw the lines on the graph.
- Write about your discoveries.
- Compare the dates of coins of different values.

7 LINE GRAPHS

A **line graph** is often used to show changes or trends over a period of time – for example, with temperature.
In these cases the horizontal axis is 'time'.

Imagine that the data is represented by a bar-line graph, then the tops of the lines in the graph joined by a series of straight lines, and the bar-lines removed. The result is a line graph.

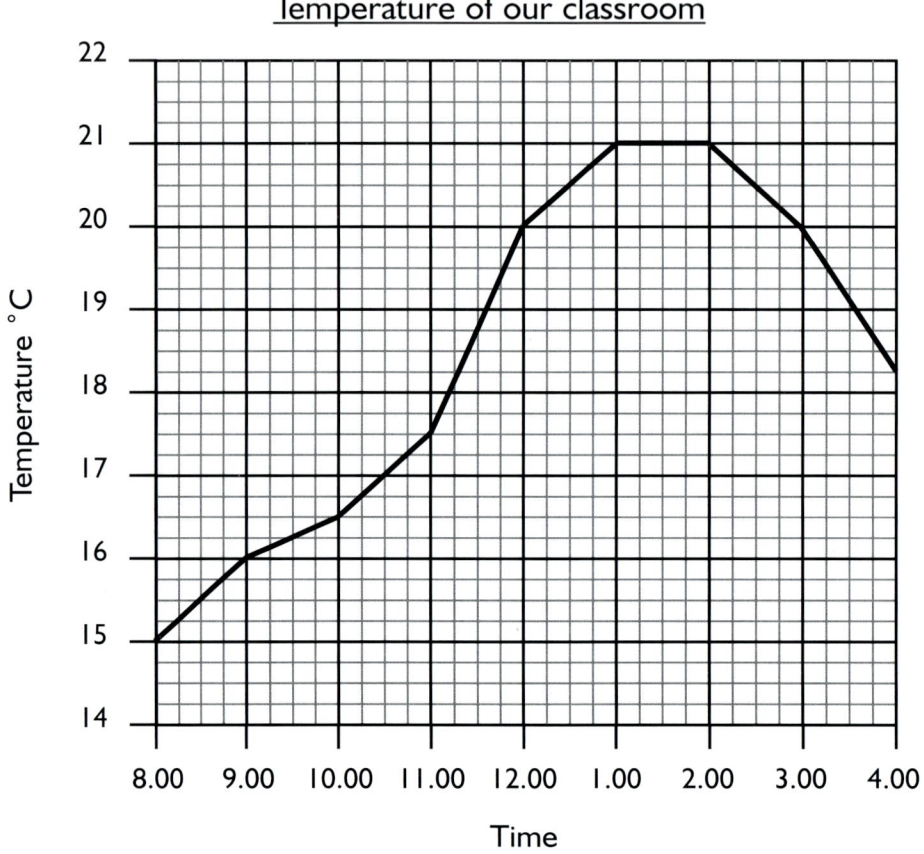

Temperature of our classroom

TO DO:

Use the graph to answer these questions.
• What was the temperature at 9.00, and at 2.00?
• At what times was the temperature 20°C?
• What was the change in temperature between 10.00 and 11.00?
• At what time was the classroom the hottest?

We can draw a line graph to show a journey.

The vertical axis shows the distance from the start in miles. The horizontal axis shows the time of the day. These graphs are often called **distance–time graphs**.

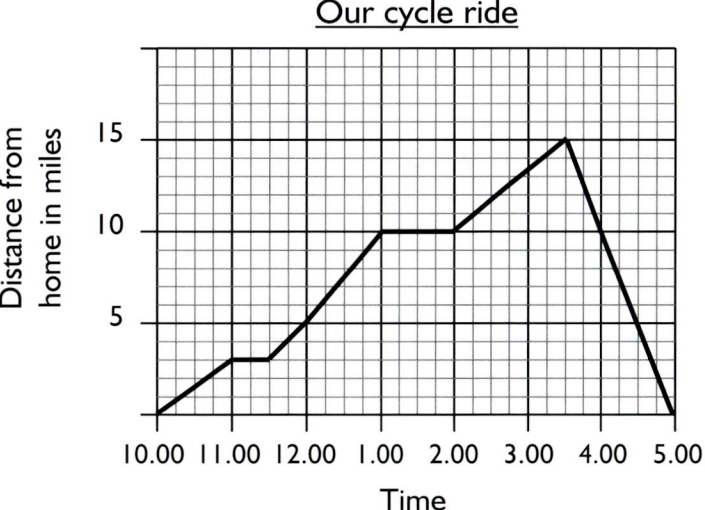

Our cycle ride

The graph shows that the cycle ride started at 10 o'clock and then 3 miles were travelled in the first hour. Between 11.00 and 11.30, the horizontal line shows that no miles were cycled and that the cyclists stopped for a rest. The journey then continued for 2 miles in the next half an hour.

TO DO:

Describe the rest of the journey.

CHALLENGE:

Invent a story to match this graph.

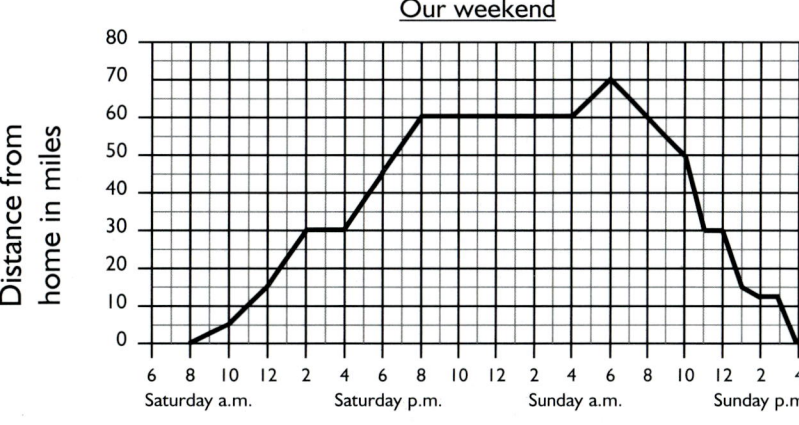

Our weekend

8 PIE CHARTS

A **pie chart** is another example of a picture which compares data. It shows how the total amount of something is shared out. It is a circle (the pie) with different-sized **sectors** (the slices of the pie).

Here is a pie chart which shows how Jenny spends her day.

<u>How Jenny spends her day</u>

Activity	Hours
Sleeping	8
Eating	2
Playing	3
Watching TV	4
At school	7

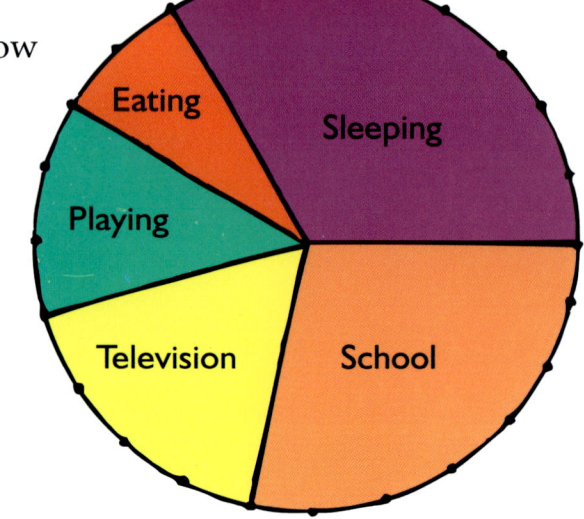

Of the 24 hours in the day, Jenny spends 3 hours playing. This is one eighth of her day. So the slice of the pie which represents her 'playing' is one-eighth of the whole pie. She spends 8 hours sleeping, which is one-third of her day. So the slice of the pie which represents her 'sleeping' is one-third of the whole pie.

The size of each sector (slice) is a fraction of the whole circle which matches the fraction of time spent on each activity in Jenny's day.

TO DO:

- Draw your own pie chart to show how you spent yesterday.
- Start by writing the data in a table.
- Draw a circle with a circular **protractor**, and divide the boundary of the circle into 24 equal divisions, one for each hour of the day. To do this, you need to make a mark for every 15°, since 360° ÷ 24 is 15°.
- Then use the data in your table to draw the different-sized sectors. Label each sector.

Data can be collected by asking people questions and recording their answers.

Gary is collecting data to find out twelve classmates' favourite sports.
This pie chart shows the results.

Gary's classmates' favourite sports

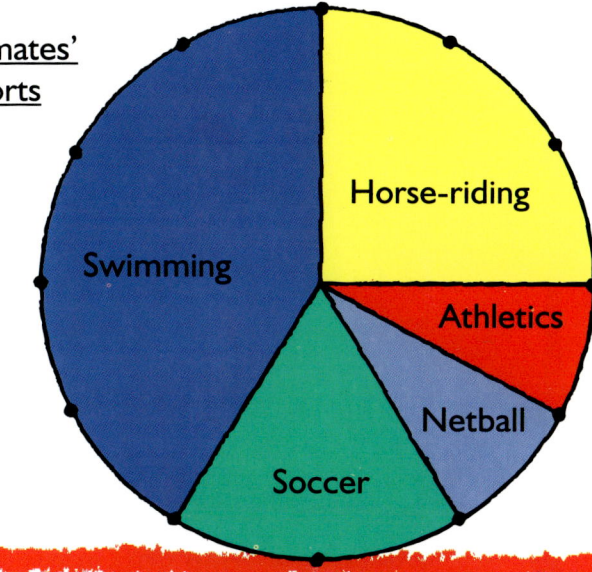

 **CHALLENGE:**

What does the pie chart tell you about how Gary's classmates voted?

 **TO DO:**

- Draw two circles and divide the boundary of each into 18 equal parts (by drawing round a circular **protractor** and marking every 20°).
- Take a pack of playing cards, shuffle them and deal out eighteen.
- On the first circle, draw a pie chart to show how many cards you have of each suit: hearts, clubs, diamonds and spades.
- On the second circle, draw a pie chart to show how many picture cards, odd-numbered and even-numbered cards you have.
- Label each chart.

9 CONVERSION GRAPHS

A **conversion graph** is a line graph used to convert from one unit of measurement to another unit of measurement.

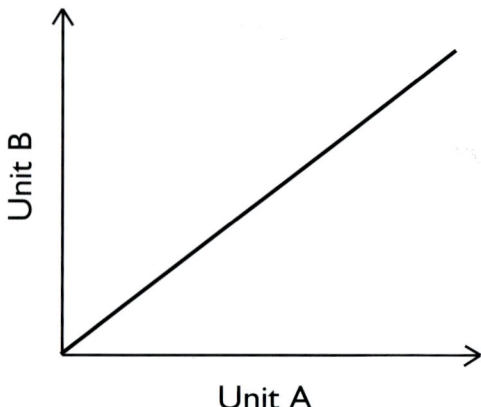

Long distances are sometimes measured in miles, and sometimes measured in kilometres.

80 kilometres is the same distance as 50 miles.

If you are used to working with miles, then you may need to convert distances from kilometres to miles. If you are used to working with kilometres, then you may need to convert distances from miles to kilometres.

A conversion graph will help you do both of these.

Conversion graph for miles and kilometres

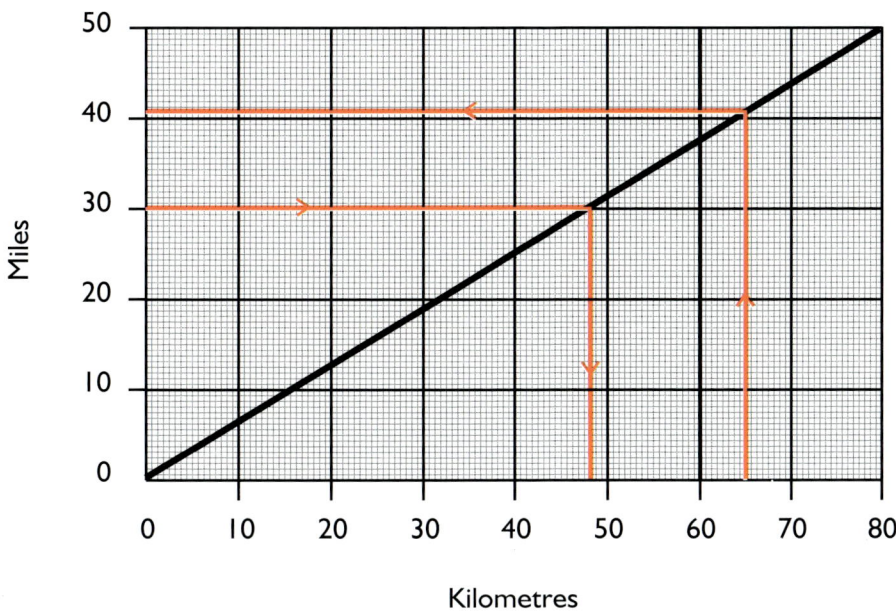

The graph shows you two conversions:
- the number of kilometres in 30 miles – 48 kilometres
- the number of miles in 65 kilometres – about 40 miles.

TO DO:

- Use the graph above to find how many kilometres there are in 20 miles and 35 miles.
- Use the graph to find how many miles there are in 20 kilometres and 55 kilometres.

CHALLENGE:

Here is a different conversion graph, to convert from one currency (pounds) to another currency (US dollars).

- Convert £5 into dollars.
- Convert $10 into pounds.

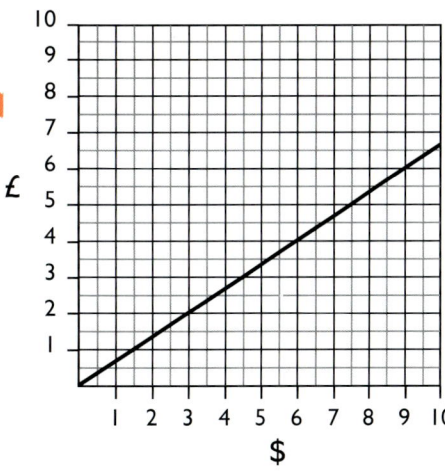

10 VENN DIAGRAMS

A **Venn diagram** is a chart which is used to sort data.
All Venn diagrams start with a rectangle, inside which are
a number of circles which sometimes cross each other
(intersect). Some examples of Venn diagrams are shown in
the diagram below.

 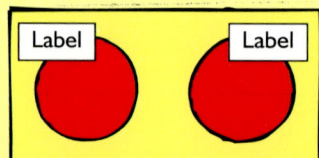 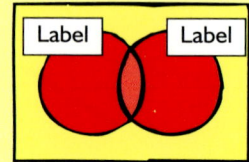 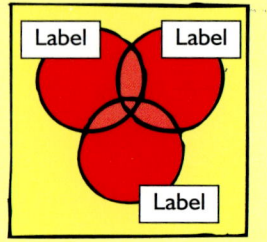

Each circle represents a set of objects or numbers,
depending on what sort of data is being sorted. Each
circle should be clearly labelled.

Instead of drawing the Venn diagram, the rectangle can be
represented by a table-top and the circles by hoops.
Objects can then be sorted on to the table.

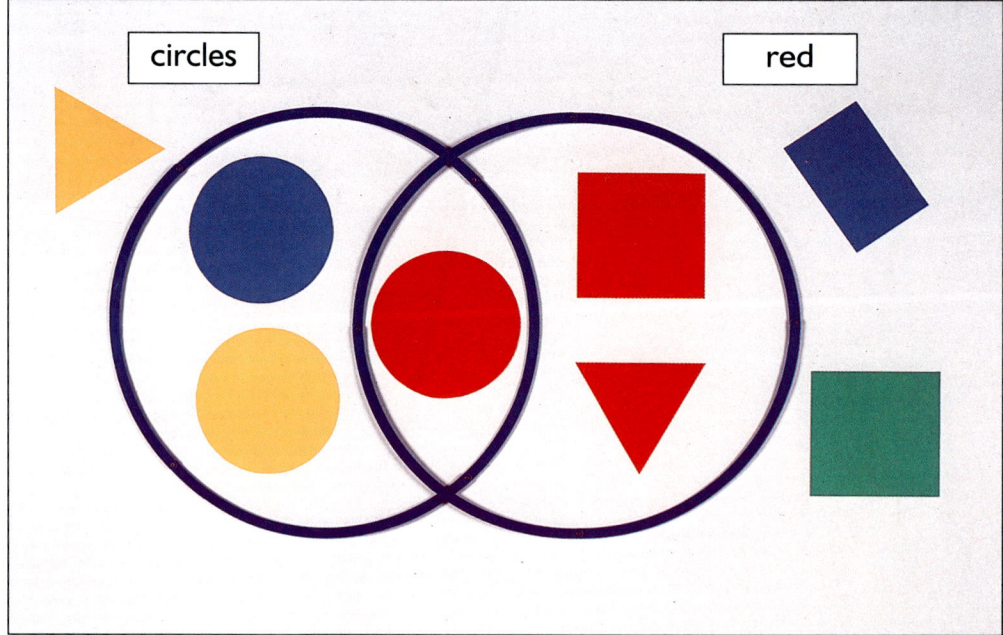

Where should a red rectangle go?
What about a blue triangle?

When the shapes have been sorted into the Venn diagram, each region contains a particular collection of shapes which can be clearly described.

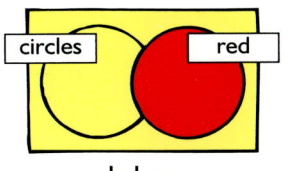

red shapes

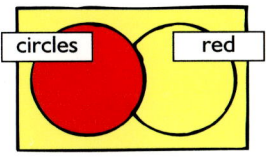

circles

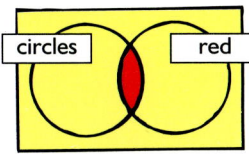

red circles

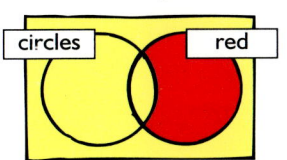

red shapes which are not circles

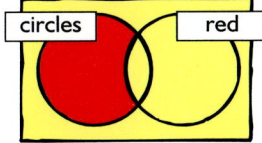

circles which are not red

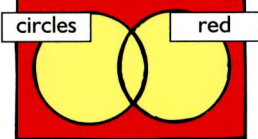

shapes which are neither circles nor red

CHALLENGE:

This Venn diagram has three circles, one for each of these sets of numbers: odd numbers, numbers more than 6, and numbers less than 12.

Two numbers are in the wrong place. What are they and where should they go?

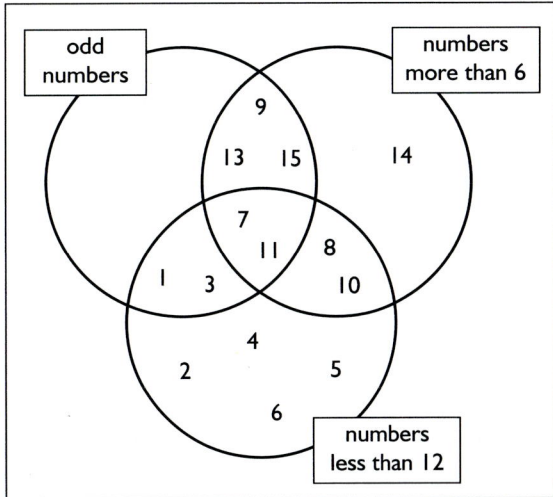

TO DO:

You need a pack of playing cards.
- Draw three large intersecting circles inside a rectangle, like the Venn diagram above.
- Label the circles 'red', 'hearts', and 'picture cards'.
- Sort the playing cards into their correct positions on the Venn diagram.
- Try inventing your own labels for the three circles, then sorting the cards.

11 DATABASES

A **database** is a collection of lots of different data about the same subjects.

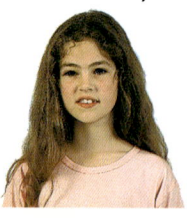

Claire
aged 8 years
1 brother
2 sisters
blue eyes
red hair

Siobhan
aged 10 years
0 brothers
1 sister
blue eyes
blonde hair

Bejal
aged 11 years
1 brother
0 sisters
brown eyes
black hair

Jack
aged 10 years
1 brother
3 sisters
brown eyes
brown hair

All of this information can be put into a table. This table is called a database.

Name	Claire	Siobhan	Bejal	Jack
Age	8	10	11	10
Brothers	1	0	1	1
Sisters	2	1	0	3
Eye colour	blue	blue	brown	brown
Hair colour	red	blonde	black	brown

The database makes it easier to search for and find information.

TO DO:

Use the table to answer the following questions.
- How many of the children have blue eyes?
- How many are aged 9 or more?
- How many have no brothers?
- How many have more sisters than brothers?
- Who has brown eyes and brown hair?

- Create a list of your own personal facts. Here are some examples of data which you might record.

Name	Height	House number	Favourite drink
Brothers	Weight	Phone number	Favourite colour
Sisters	Hair colour	Lucky number	Favourite TV programme
Age	Eye colour	Favourite food	Favourite music

- Collect data from your friends and build a table for your own database.

A football league table is another example of a database. The data in the table summarizes current performance of each team in the league.

Team	Matches won	Matches drawn	Matches lost	Goals scored	Goals conceded	Points
Athletico	22	12	7	74	42	78
Dynamo	23	8	9	73	32	77
United	20	12	6	70	30	72
Racing Club	18	15	7	62	35	69
Olympic	19	11	11	61	48	68
White Star	20	6	13	69	63	66

The teams are awarded 3 points for a win, 1 point for a draw, and 0 points for a loss.

CHALLENGE:

Look at the league table above.
- Which team has scored the fewest goals? Which the most?
- Which team has drawn the most matches? Which the least?
- Which team has the largest difference between goals scored and goals conceded?
- Suppose there were only 2 points for a win instead of 3. How many points would each team have? Would they be in the same positions in the league?

TO DO:

Create a database to show the results of these matches in the form of a league table:

Kickton City 2 Shotville 1

Shotville 2 Hardtackle Utd 3

Passwell Town 1 Fastmen Rovers 0

Fastmen Rovers 0 Kickton City 0

Hardtackle Utd 3 Fastmen Rovers 3

Kickton City 1 Passwell Town 1

Shotville 4 Fastmen Rovers 5

Hardtackle Utd 3 Kickton City 1

Passwell Town 1 Shotville 4

Hardtackle Utd 2 Passwell Town 2

DECISION TREES

A **decision tree** can be used to sort objects. It is called a tree because it has several branches (or routes). At the start of a pair of branches is a question to which the answer is either 'Yes' or 'No'. This is where you have to make the decisions.

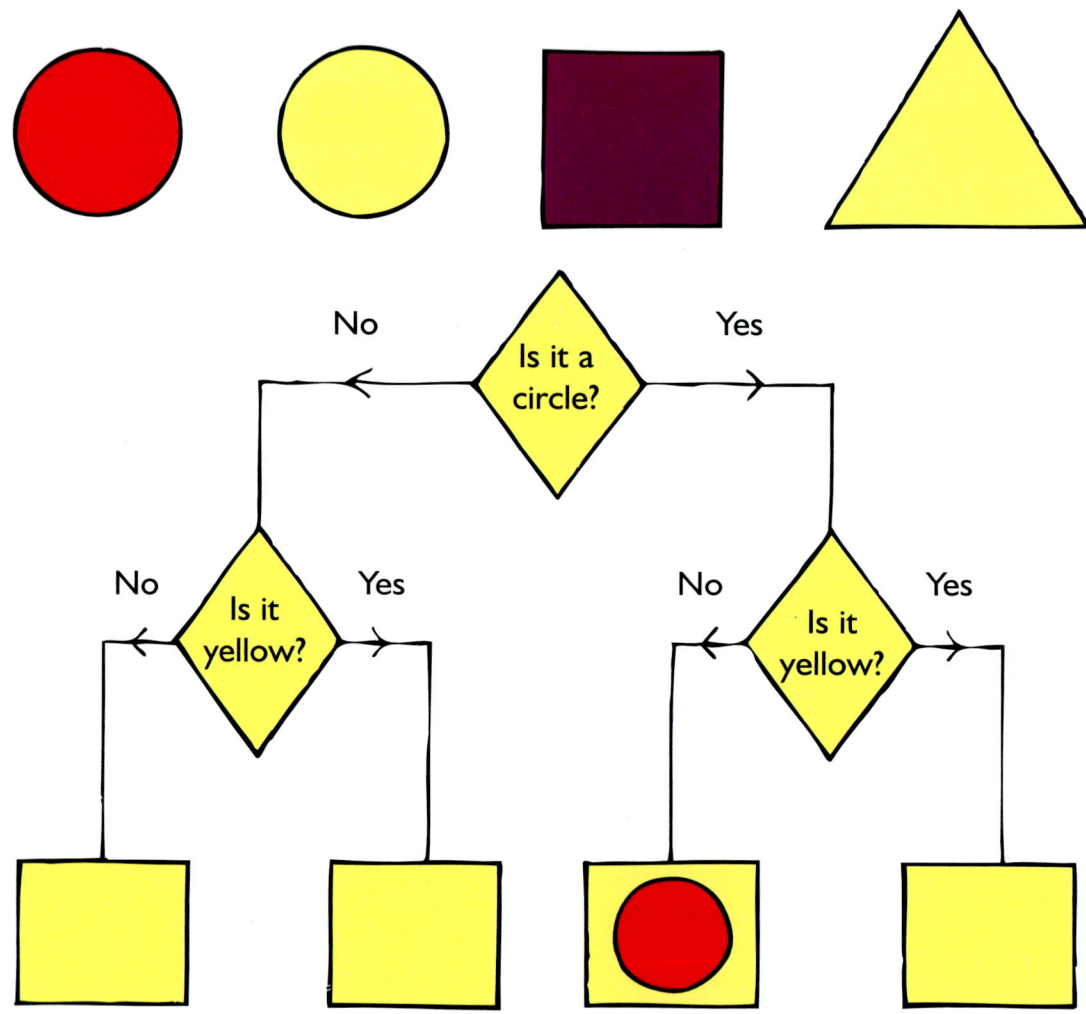

This decision tree will sort the shapes.
Start by sorting the red circle.
The answer to the first question, 'Is it a circle?' is 'Yes'.
Follow the branch until you come to the next question, 'Is it yellow?' The answer is 'No'.

TO DO:

Sort the other shapes to decide where they go.

Playing cards can be sorted in many different ways.

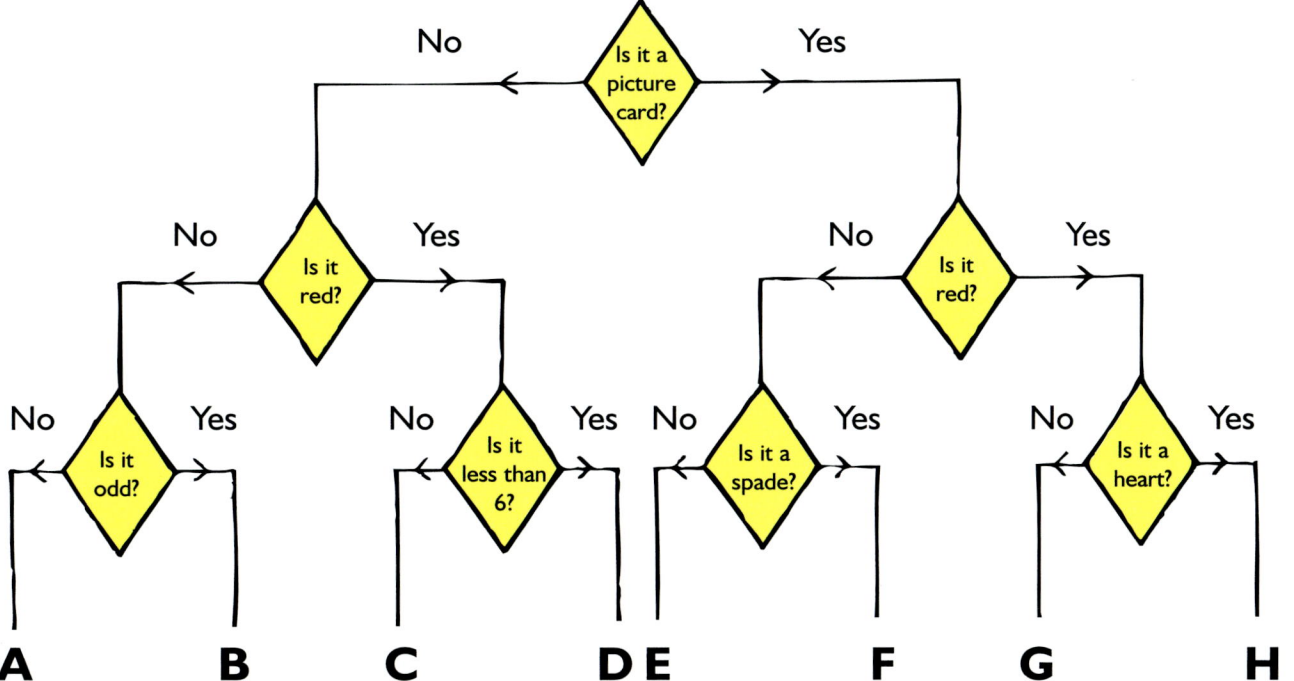

CHALLENGE:

Take a pack of playing cards, shuffle them, then try each in turn using the decision tree. Find out how many cards will finish in each of the eight positions labelled A to H.

TO DO:

Invent your own decision tree to sort a pack of playing cards, then try it out.

13 AVERAGES

An **average** of a set of numbers is a middle number about which they are centred. For example, the average of the numbers 27, 29, 35, 38, 24, 30 and 34 is 31.

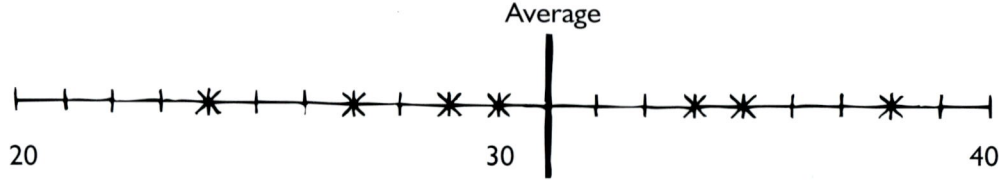

To find the average of a set of numbers, you find their total and divide this by the number in the set. The set of numbers 3, 4 and 8, for example, has a total of 15 and there are three of them. So their average is 15 ÷ 3, which is 5.

The average of a set of numbers is often not a whole number. The set of numbers 7, 2, 6 and 3, for example, has a total of 18 and there are four of them. So their average'is' 18 ÷ 4, which is 4.5.

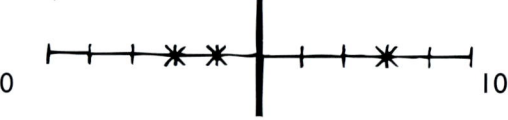

This average is sometimes called the **mean**.

TO DO:

Play the average dice game

You need five dice.
- Take turns to throw the dice. You may choose to throw 2, 3, 4 or 5 dice.
- Your score is the average dice number to the nearest whole number. For example, if you choose 3 dice and throw:

 then the total is 11. There are 3 dice, so the average is 11 ÷ 3, which is 4 to the nearest whole number. This is your score.

- The player with the highest average wins the round.
- Check each other's scores.
- The winner is the first to win six rounds.

The average
contents of a box
of paperclips is
stated on the box.

'Average Contents: 40' means that some boxes may have a
few less than 40 paperclips, and some a few more, but
overall the number of paperclips in a box will be centred
around 40.

TO DO:

Find the average number of letters in a word

Alice was beginning to get very tired of sitting by her sister on the
bank, and of having nothing to do: once or twice she had peeped
into the book her sister was reading, but it had no pictures or
conversations in it, 'and what is the use of a book,' thought Alice,
'without pictures or conversation?' So she was considering in her
own mind (as well as she could, for the hot day made her feel
very sleepy and stupid), whether the pleasure of making a daisy
chain would be worth the trouble of getting up and picking the ...

Here are a hundred words from a book.
The number of letters in the first few words are:
5, 3, 9, 2, 3. Use a tally chart to record the number of letters
in all the hundred words, and use this to find the average
number of letters in a word.

CHALLENGE:

Find the average number of letters in another book or
newspaper. Compare this average with the average found
above.

14 PROGRAMMES

Useful data is presented in the form of a programme or schedule. If you are planning to visit the cinema, then you need to know what films are showing, and at what time they start.

CHALLENGE:

Study the cinema programme. The film 'Little Men' is shown three times on a Friday and four times on every other day. The film lasts for 2 hours and 15 minutes.

- How much time is there between when the film ends and when it starts again?
- How many times is it shown in a week?
- How many times are the other films shown in the week?
- How many film showings are there on each day of the week? How many in the week altogether?
- Which is the longest film?
- Which is the shortest?

Quiet and Quieter (12) 2 hrs
(two screens)

Fri/Sat	1.00 1.45 3.30 4.15 6.00 7.00 8.30 9.40 10.55 12.00
Sun–Thurs	11.15 1.00 1.45 3.30 4.15 6.00 7.00 8.30 9.40

Little Men (U) 2 hrs 15 mins

Fri	2.00 5.00 8.00
Sat–Thurs	11.30 2.00 5.00 8.00

Enclosure (18) 2 hrs 20 mins

Fri/Sat	12.00 3.20 6.20 9.15 11.50
Sun–Thurs	12.00 3.20 6.20 9.15

Small Adventure (15) 2 hrs 5 mins

Fri/Sat	2.10 4.40 7.20 9.50
Sun–Thurs	11.45 2.10 4.40 7.20 9.50

Terminal Speed (15) 2 hrs

Fri/Sat	7.30 10.00 12.10
Sun–Thurs	1.45 12.45 3.00

1001 Puppies (U) 1 hr 35 mins

Daily	10.15 12.30 2.45 5.10

Near to Work (U) 1 hr 35 mins

Fri/Sat	12.45 3.00
Sun–Thurs	10.45 12.45 3.00

Kindness of Jim (PG) 2 hrs 10 mins

Daily	12.10 2.30 5.30 8.15

TO DO:

Find a cinema programme from the local newspaper and analyse the data.

In the same way, if you are going to watch television you need to see the programmes on offer, the start times, and when they finish.

CHALLENGE:

Look at the television schedule below.
- Which programme starts at 8.55?
- Which programme finishes at 7.10?
- Which programme follows after 'Private Lives'?
- How long does 'Felix the Cat' last?
- How many programmes could you watch between 5.00 and 8.00 in the evening?
- Which programmes last for exactly half an hour?
- How many last for longer than half an hour?
- How many last for less then half an hour?
- Which programme is the longest?
- How long does each programme last?

Saturday

7.25	News; Weather	5.15	News; Weather
7.30	Pingo (rept)	5.25	Regional News
7.35	Felix the Cat	5.30	Cartoon
7.50	The Five Musketeers	5.50	Quiz Show
8.15	Joketime	6.20	Batman
8.35	The Baboons	7.10	Private Lives
9.00	Playtime	7.50	Lottery
12.12	Saturday Sport	8.05	Nasties
	12.20 *Racing*	8.55	News; Sport; Weather
	12.30 *Snooker*	9.15	Gunsmoke
	12.40 *Football*	10.00	Bob Smith Show
	1.00 *News*	10.40	Match of the Day
	1.05 *Athletics*	11.40	Comedians
	3.05 *Racing*	12.10	Film
	4.00 *Golf*	1.50	News; Weather
	4.40 *Results*		

 **DISTANCE CHARTS**

A **distance chart** allows you to find the distance between two towns or cities.

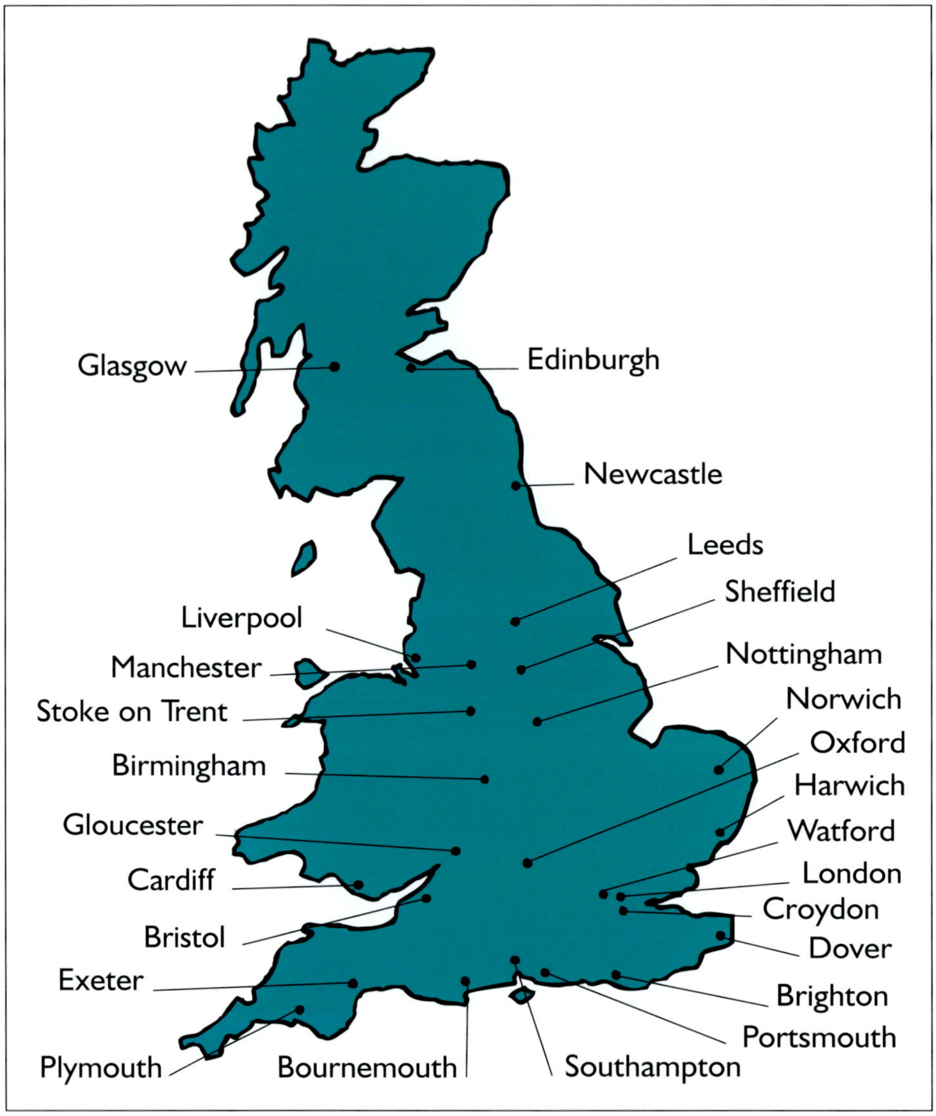

Glasgow

Edinburgh

Newcastle

Leeds

Sheffield

Liverpool

Nottingham

Manchester

Norwich

Stoke on Trent

Oxford

Birmingham

Harwich

Gloucester

Watford

Cardiff

London

Croydon

Bristol

Dover

Exeter

Brighton

Portsmouth

Plymouth

Bournemouth

Southampton

If you are going on a journey from one town to another, then you need to know how far it is, and how long it will take.

 **CHALLENGE:**

Do you think it is farther from Oxford to Leeds, than it is from Gloucester to Harwich?
A distance chart will help you find out.

Distance chart in miles

London																									
120	Birmingham																								
104	145	Bournemouth																							
57	183	94	Brighton																						
118	89	75	167	Bristol																					
153	110	126	203	47	Cardiff																				
14	157	117	43	142	178	Croyden																			
79	202	180	83	205	240	78	Dover																		
405	291	438	468	373	394	442	487	Edinburgh																	
198	162	84	171	82	120	191	254	446	Exeter																
402	288	435	464	370	390	439	484	46	443	Glasgow															
105	54	104	162	35	57	137	199	338	108	335	Gloucester														
82	168	201	132	215	251	90	130	418	295	425	190	Harwich													
197	133	256	260	215	236	234	279	210	288	247	180	220	Leeds												
213	98	246	275	180	202	250	295	222	253	219	145	261	73	Liverpool											
286	204	345	347	298	319	322	350	112	371	149	263	302	94	184	Newcastle										
115	163	230	177	244	280	134	175	374	324	380	184	74	175	219	257	Norwich									
129	57	188	191	151	172	166	211	277	224	283	116	162	72	110	160	122	Nottingham								
57	65	93	108	73	108	83	145	358	150	354	48	142	164	165	252	144	96	Oxford							
238	202	126	213	123	160	232	294	486	44	483	148	330	328	294	411	336	264	193	Plymouth						
72	148	51	50	96	142	78	140	451	128	448	117	174	247	258	335	204	179	83	171	Portsmouth					
169	89	228	232	182	203	206	251	250	255	257	147	196	35	79	134	152	44	136	295	219	Sheffield				
77	130	32	63	76	122	90	152	433	109	429	99	174	229	240	317	202	161	65	150	19	201	Southampton			
199	85	233	262	167	188	237	281	224	241	221	132	223	58	41	169	178	66	152	280	245	38	227	Stockport		
159	45	194	222	128	150	196	241	249	201	246	93	208	91	56	202	172	51	112	241	206	52	188	43	Stoke on Trent	
21	102	110	82	125	160	57	114	387	205	384	99	93	179	195	267	123	111	51	244	84	151	82	181	141	Watford

To find the distance between two towns, look at the square in which the row and column meet. For example, to find the distance between Cardiff and Norwich, look down the Cardiff column, until you meet the Norwich row. The meeting square shows you that the two towns are 280 miles apart.

TO DO:

- Find five pairs of towns which are more than 300 miles apart.
- Find five pairs of towns which are less than 50 miles apart.

16 TIMETABLES

A **timetable** provides data to help you plan a journey.
This is a timetable for the number 51 bus.

BANKS BROOK - CITY MOOR - CROSSTOWN - HILLTOWN - GOLDHILL 51

MONDAY to FRIDAY

								0807								
Banks Brook,																
City Moor, City Moor Hospital,	0522	0612	0649	0719	0737	0747	0757	0807	0817	0827	0837	0847	0857	-	0907	0917
Crosstown, Sandy Road,	0530	0620	0657	0727	0745	0755	0805	0815	0825	0835	0845	0855	0905	-	0915	0925
Hilltown, Leopold Street,	0545	0635	0712	0742	0800	0810	0820	0830	0840	0850	0900	0910	0920	0925	0930	0940
Goldhill, Leighton Road	0612	0702	0739	0809	0827	0837	0847	0857	0907	0917	0927	0937	0947	-	0957	1007

Banks Brook,														
City Moor, City Moor Hospital,	0927	0937	0947	0957	then at	07	17	27	37	47	57		1457	1507	1514
Crosstown, Sandy Road,	0935	0945	0955	1005	these	15	25	35	45	55	05	until	1505	1515	1522
Hilltown, Leopold Street,	0950	1000	1010	1020	minutes	30	40	50	00	10	20		1520	1530	1537
Goldhill, Leighton Road	1017	1027	1037	1047	past hour	57	07	17	27	37	47		1547	1557	1604

										1634							
Banks Brook,																	
City Moor, City Moor Hospital,	1522	1529	1537	1544	1552	1559	1607	1614	1622	1629	1637	1644	1652	1659	1707	1714
Crosstown, Sandy Road,	1530	1537	1545	1552	1600	1607	1615	1622	1630	1637	1645	1652	1700	1707	1715	1722
Hilltown, Leopold Street,	1545	1552	1600	1607	1615	1622	1630	1637	1645	1652	1700	1707	1715	1722	1730	1737
Goldhill, Leighton Road	1612	1619	1627	1634	1642	1649	1657	1704	1712	1719	1727	1734	1742	1749	1757	1804

Banks Brook,											
City Moor, City Moor Hospital,	1722	1729	1737	1754	then at	14	34	54		2254	2314
Crosstown, Sandy Road,	1730	1737	1745	1801	these	21	41	01	until	2301	2321
Hilltown, Leopold Street,	1745	1752	1800	1815	minutes	35	55	15		2315	2335
Goldhill, Leighton Road	1812	1819	1827	1840	past hour	00	20	40		2340	-

To read the times of the journey for a bus, you need to look at a column of times.
The first column looks like this:

Banks Brook	–
City Moor	5.22
Crosstown	5.30
Hilltown	5.45
Goldhill	6.12

This is the first bus in the morning. It leaves City Moor at 5.22 in the morning, and arrives at Goldhill at 6.12.
How long does the whole journey take?

As you look along the different columns, you will see that there are lots of buses travelling from City Moor to Goldhill. The last bus leaves at 10.54 pm and arrives at 11.40 pm.

CHALLENGE:

Can you work out how long the bus takes to travel from:
• City Moor to Crosstown
• Crosstown to Hilltown
• Crosstown to Goldhill?

After 10.00 in the morning the buses run regularly, so the timetable has been shortened.
• How many buses run each hour in the middle of the day?
• How often do they run?

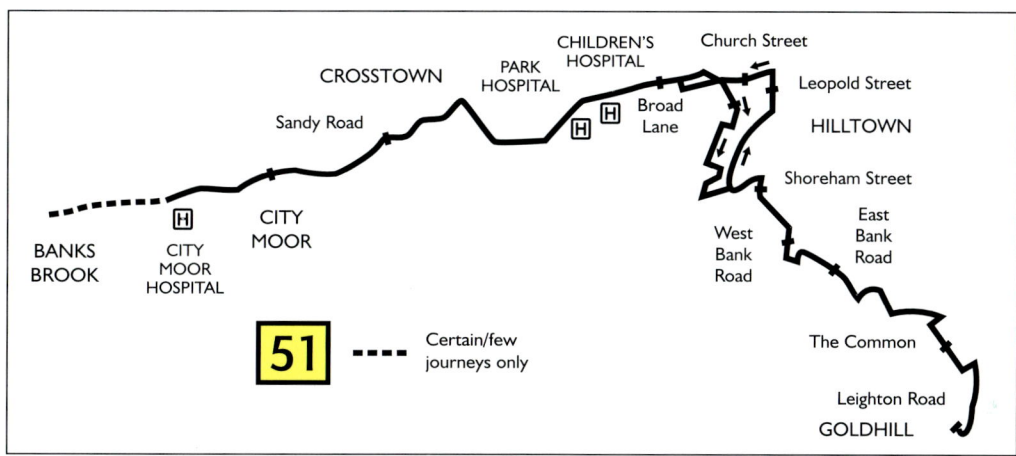

The timetable does not give the time of arrival at each stop, so passengers need to estimate the times of arrival at bus stops between those listed on the timetable. For example, if you want to catch a bus from Shoreham Street, then you can look at the times when the bus arrives at Hilltown, Leopold Street, and assume that the bus will arrive at Shoreham Street a few minutes later.

You will see that there are two buses each day which start before City Moor, at Banks Brook.
At what times do these buses start?

CHALLENGE:

Can you work out how many buses travel each day on route 51 (including the two from Banks Brook)?

17 CHANCE

If something may or may not happen, you sometimes need to decide how likely it is to happen. This is called the **chance** that it will happen.

- *Will it rain tomorrow?*
- *Will you grow up to be taller than your mother?*
- *Will you become famous?*

If it has a good chance of happening, it is more likely to happen than not. If it has a poor chance of happening, it is less likely to happen than not.

If it has an **even chance** of happening, it is as likely to happen as not. This even chance is sometimes called a fifty-fifty chance.

If it has no chance of happening, it is impossible. If it is sure to happen, it is certain.

TO DO:

What are the chances of these things happening to you?

Use the words 'impossible', 'poor chance', 'even chance', 'good chance' and 'certain'.

I will visit Australia	I will travel in a rocket	It will rain this year	I will turn into a frog
It will get dark tonight	I will get younger everyday	I will one day be married	I will learn to drive a car
I will live for 100 years	I will become a teacher	I will appear on television	I will grow to be 6 feet tall

These different chances can be shown in order on a line.

If this line is matched to a number line from 0 to 1, then the positions of the chances on the line are called **probabilities**.

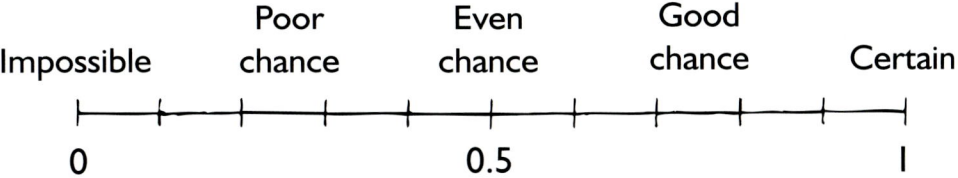

- If something is impossible, its probability is 0.
- If something is certain, its probability is 1.
- If something has a fifty-fifty chance of happening, its probability is $\frac{1}{2}$.
- If something has a good chance of happening, its probability is between $\frac{1}{2}$ and 1.

A probability is a number between 0 and 1. The probability that something will happen is a measure of its chance of happening.

TO DO:

Play the probability game

- Start by making twelve probability cards, three of each of these:

| Probability is 0 | Probability is between 0 and $\frac{1}{2}$ | Probability is between $\frac{1}{2}$ and 1 | Probability is 1 |

- Shuffle the cards and place them in a face-down pile.
- Take turns to reveal the top card: for example,

| Probability is between $\frac{1}{2}$ and 1 |

and throw a dice: for example,

- Use the key on the right to match the dice number to a time: 'tomorrow afternoon'.

- Invent an outcome for your partner to match the probability card and the time: for example, 'Tomorrow afternoon you will smile.'
- Discuss each other's statements to see if you agree.

Dice key

next Saturday

this evening

tomorrow afternoon

in the year 2000

next week

next Christmas

18 ROLLING DICE

A six-faced dice is used in many games of chance, because when it is rolled, each face has an equal chance of appearing.

This tally chart shows the results of rolling a dice 60 times.

Dice number	Tallies		Frequency
1	++++ IIII		9
2	++++ III		8
3	++++ ++++	II	12
4	++++ ++++		10
5	++++ ++++	III	13
6	++++ III		8

Since each face of the dice has an equal chance of appearing, it is expected that approximately one sixth of all the throws will show '1', another sixth will show '2', and so on. As the dice was rolled 60 times, we expect approximately ten of each number. We expect them to be near to ten, but very rarely will they be exactly ten.

We say: 'The probability of throwing a 1 is $\frac{1}{6}$.'
So is the probability of throwing a '2', a '3', a '4', a '5' and a '6'.

TO DO:

- Roll a dice 120 times, but first decide how many times you expect each number to appear.
- Record your results in a table, and compare the totals with your expectations.

TO DO:

Play the two-dice horse-race game

You need a red and a green dice and some counters.

- Start by drawing a large version of the race track below.
- Place a counter (the horse) at the start of each track.
- Throw both dice, find the total, and move the matching horse forward one space.
- Continue throwing the dice, and moving a matching horse one space.
- The winning horse is the first to reach 'Finish'.

Play several games, and see which are the winning horses. Do some horses have better chances of winning than others?

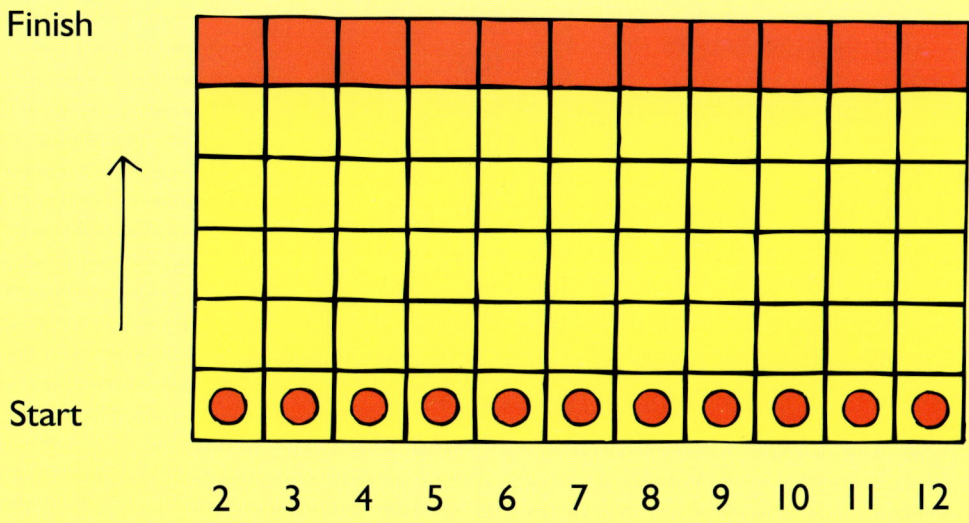

Finish

Start

2   3   4   5   6   7   8   9   10   11   12

19 TOSSING COINS

When a coin is tossed, it can land either heads or tails. Each is equally likely. Each has an even chance. Each has a probability of $\frac{1}{2}$.

When two coins are tossed, sometimes both will show heads, sometimes both will show tails, and sometimes there will be one head and one tail.

TO DO:

You need a large coin and a small coin.
- Toss the large coin first, then the small coin.
- Do this twenty times, and see how may of the tosses result in 'both heads', how many in 'both tails' and how many in 'one of each'.

The different possible happenings are as follows:

Large coin	Small coin	
Head	Head	(HH)
Head	Tail	(HT)
Tail	Head	(TH)
Tail	Tail	(TT)

All of these are equally likely. So approximately one-quarter of the tosses should show two heads, another quarter should show two tails, and a half of the tosses should show a head and a tail. Compare this with the results of your twenty throws.

The probability of tossing two heads is $\frac{1}{4}$.
The probability of tossing two tails is $\frac{1}{4}$.
The probability of tossing a head and a tail is $\frac{1}{2}$.

When three coins are tossed, the different possible happenings or outcomes can be shown by a **tree diagram**. First a small coin is tossed, then a medium-sized coin and finally a large coin.

Tree diagram for tossing 3 coins

Small coin	Medium-sized coin	Large coin	Outcome
	H	H	HHH
		T	HHT
H	T	H	HTH
		T	HTT
	H	H	THH
T		T	THT
	T	H	TTH
		T	TTT

There are eight different paths from left to right along the 'branches of the tree'.

Each path shows one of the eight different possible outcomes.
Each outcome is equally likely, and so each has a probability of $\frac{1}{8}$.

One of the eight outcomes results in no heads at all – TTT.
Three of these outcomes result in one head – HTT, THT, TTH.
Three result in two heads – HTH, HHT, THH.
One results in three heads – HHH.

These can lead to a probability table.

Probability table for tossing 3 coins	
Number of heads	Probability
0	$\frac{1}{8}$
1	$\frac{3}{8}$
2	$\frac{3}{8}$
3	$\frac{1}{8}$

TO DO:

You need a small, a medium-sized and a large coin.
- Toss the three coins 24 times altogether.
- Count the number of times you throw 0 heads, 1 head, 2 heads and 3 heads.
- Compare your results with your expectations.

CHOICES

Sometimes, when we have a choice, we need to know how many different possible choices can be made.

Sandwiches

Fillings: Ham
　　　　　Cheese and tomato
　　　　　Chicken

Bread:　White
　　　　　Brown

One possible choice is a ham sandwich in white bread. Another is a chicken sandwich in brown bread.

There are six different possible choices altogether. What are the other four?

Menu

Fish fingers or sausages

Chips, boiled potatoes or mashed potatoes

Beans or peas

Karen chose sausages, mashed potatoes and beans.
John chose fish fingers, chips and peas.
What would you choose?

Can you find twelve different choices altogether?

When choosing a sandwich there were 3 fillings and 2 types of bread, giving 3 x 2 = 6 different choices.
When choosing from the menu, there were 2 main dishes, 3 types of potato and 2 vegetables, giving 2 x 3 x 2 = 12 different choices.

If you are organising a tournament, you need to know how many matches need to be scheduled.

There is going to be a draughts competition for five players. Each player must play every other player, and we need to know how many matches will be played altogether.

If the five players are labelled A, B, C, D and E, we can work systematically by:
- listing all of A's matches
- then listing B's remaining matches, not counting the match against A
- then listing C's remaining matches, and then D's.

All of E's matches should now be listed.

A v B	B v D	C v D	D v E
A v C	B v D	C v E	
A v D	B v E		
A v E			

There are 10 matches altogether. Check that none are missing.

Each match can be shown by a line joining two points on this diagram. Check that there are 10.

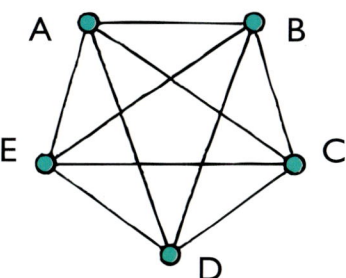

The matches can also be shown on a chart like this. The shaded squares shown the 10 matches.

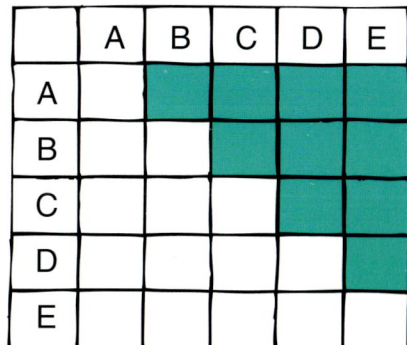

CHALLENGE:

- Four people are in a room and each must shake the hand of the other three people. Find how many handshakes there will be altogether.
- Extend the problem to five people and then six.
- Can you spot a pattern in the number of handshakes each time?

GLOSSARY

average   A number about which numbered **data** is centred.

axes   The two lines which explain the **data** being shown on a graph. One line is **horizontal** and one is **vertical**.

bar graph or bar chart   A graph which uses bars to show the **data**. The heights of the bars show frequencies.

bar-line graph   A graph in which the bars have been replaced by lines.

block graph   A graph which uses blocks to show the **data**.

chance   The chance of something happening is how likely it is to happen.

conversion graph   A line graph which is used to convert from one unit of measurement to another.

database   A large collection of **data** which can be sorted in different ways.

data   Facts and information about something.

data collection sheet   A sheet for recording **data**.

data processing   Simplifying and ordering recorded **data**.

data representation   Showing the **data** in an easy-to-read way.

decision tree   A chart which has different routes (branches) to follow after answering questions.

distance chart   A chart which helps you find the distance between two places.

distance-time graph   A line graph in which the axes are distance and time.

even chance   Equally likely to happen as not to happen.

frequency   How many or how often.

grouped data   Sets of scores or observations put together into groups.

horizontal line   A straight line drawn from left to right on the paper.

interpreting data Reading and understanding **data**.

key   A guide to explain the meaning of part of a graph.

line graph	A graph with a continuous line(s) to show trends or changes.
mean	An **average**.
pictograph	A graph which uses pictures to show .
pie chart	A graph or chart which shows data by slicing a pie (circle) into slices (sectors).
probability	A measure of chance, between 0 and 1.
protractor	An instrument for measuring angles.
questionnaire	A set of questions used to collect **data**.
tally	Line or marks to score the number of times something occurs.
tally chart	A chart or table for drawing tallies.
timetable	A chart showing times for different parts of a journey or event.
tree diagram	A chart with 'branches of a tree'to show different possible happenings.
Venn diagram	A chart which sorts **data** by placing it in circles.
vertical line	A straight line drawn from the top to bottom, at right-angles to a horizontal line.

INDEX

ANSWERS

p6	Mint, 5 shoppers. Vanilla
p7	Summer 10; autumn 6; winter 3; altogether 26
p8	**To do:** 11; 4; 13
p9	**Challenge:** 30-39; 4; 31; 18
p11	**To do:** 2, brown and black; 3; 11
p13	**Challenge:** One possible solution is:

p16	**To do:** 16°C, 21°C; 12 o'clock and 3 o'clock; 1°C increase; between 1 o'clock and 2 o'clock.
p19	**Challenge:** Horseriding – 3 votes, Athletics – 1 vote, Netball – 1 vote, Soccer – 2 votes, Swimming – 5 votes.
p21	32km and 54km; 12.5 miles and 34.5 miles. **Challenge:** $7.50; £6,50
p23	5 and 9 are wrongly placed.
p24	2; 3; 1; 3; Jack.
p25	**Challenge:** Olympic; Athletico Racing Club; White Star; Dynamo; Athletico, 56pts; Dynamo, 54pts, United, 52pts; Racing Club, 51pts; Olympic, 49pts; White Star, 46pts
p29	Average of 4 letters in a word.

p30   45 mins; 27; Quiet and Quieter, 65; Little Men, 27; Enclosure, 30; Small Adventures, 33; Terminal Speed, 21; 1001 Puppies, 28; Near to Work, 19; Kindness of Jim, 28. 36 every day except 35 on Friday. Longest - 'Enclosure', shortest - '1001 Puppies' and 'Near to Work'.

p32   Gloucester to Harwich is farther.

p34   50 minutes.

p35   **Challenge:**
8 minutes; 15 minutes; 42 minutes; 6 buses each hour; every 10 minutes; 8.07, 16.34.

Challenge:
86 buses.

p42   Fish fingers, chips, beans. Fish fingers, boiled potatoes, beans. Fish fingers, mashed potatoes, beans. Fish fingers, chips, peas. Fish fingers, boiled potatoes, peas.
Fish fingers, mashed potatoes, peas. Sausages, chips, beans. Sausages, boiled potatoes, beans. Sausages, mashed potatoes, beans. Sausages, chips, peas. Sausages, boiled potatoes, peas. Sausages, mashed potatoes, peas.

p43   **Challenge:**
handshakes: 4 people, 6; 5 people, 10; 6 people, 15.